A SHORTER BIOETHICS

FRANCIS ETHEREDGE

Foreword by Dr. Thomas Williams

En Route Books and Media, LLC
Saint Louis, MO

En Route Books and Media, LLC
5705 Rhodes Avenue
St. Louis, MO 63109

Cover credit: Sebastian Mahfood using DALL-E

Copyright © 2025 Francis Etheredge

ISBN-13: 979-8-88870-429-5

Library of Congress Control Number:
Available at https://catalog.loc.gov

No part of this book may be reproduced, stored in a retrieval system, or transmitted in any form, or by any means, electronic, mechanical, photocopying, or otherwise, without the prior written permission of the author or the publisher.

Table of Contents

Foreword by Dr. Thomas Williams v

Introduction—*Why Am I a Writer?*, and *Three Kinds of Published Work* 1

Preamble: What is Bioethics? *Listening and being listened to; Suffering and the Search for Meaning; and The Christian Depths of Human Being* 7

Chapter One: Faith and Reason *Change and Constancy; The Contribution of Faith; and Ethics-in-Relationships* . 13

Chapter Two: The Whole Human Being and Contraception *The Sacrament of Marriage; Being Open to Life: And a Practical Spirituality* 25

Chapter Three: An Actual Beginning: Conception *Why Do We Need to Draw Upon the Existence of God; How does this help us with the question of the conception of a human person? Why God is a necessary cause of human existence* 35

Chapter Four: The Frozen Embryo Debate and *in-vitro* fertilization (IVF) *Questions that need to be asked and answers that need to be given; The present situation of a frozen human embryo; Is there a theological justification for helping the frozen child?* 45

Chapter Five: Identity; Transgenderism; and the United Nations Bureaucrats *Growing up is Growing Through; Can I Be in the Wrong Body; Ideology*............ 55

A Word About Free Speech 65

Foreword by Dr. Thomas Williams

We live in an age of unprecedented confusion concerning the nature and identity of the human person. When a U.S. Supreme Court justice can assert she is unqualified to define what a woman is because she is not a "biologist,"[1] the descent into ideologically driven ignorance is all but complete, as is the moral mayhem that necessarily follows. When the killing of a healthy unborn child is proclaimed as the pinnacle of human rights among the upper echelons of society,[2] the very idea of bioethics is stood on its head.

The moral relativism of our times proposes that there is no universal or absolute set of moral principles to which human beings must adhere. Coupled with a denial of the natural moral law, this relativism leads many to suppose that ethics is more a question of tastes, preferences, feelings, and personal beliefs than an antiquated notion of universal "moral truth." Nowhere is this more evident than in the thorny field of bioethics, which deals with the morality of behavior touching upon human life in all its stages.

---

[1] See Politico, "Blackburn to Jackson: Can you define 'the word woman'?" (March 22, 2022) at https://www.politico.com/news/2022/03/22/blackburn-jackson-define-the-word-woman-00019543.

[2] See NBC News, "Biden administration announces new abortion initiatives on Roe anniversary," at https://www.nbcnews.com/politics/joe-biden/biden-administration-announces-new-abortion-initiatives-roe-anniversar-rcna134954.

Fortunately, Francis Etheredge has provided the world with a compelling yet accessible remedy to this anthropological and moral chaos with the publication of *A Shorter Bioethics*. This handy volume offers readers answers to the crucial questions that plague modern society with the advance of biotechnology and a loss of the sense of the inviolable sacredness of human life.

As its title suggests, *A Shorter Bioethics* is a concise treatment of the central tenets of bioethics from the perspective of moral realism and the Revelation-inspired teachings of the Catholic Church. This perspective begins with an understanding of the human person as made in the image and likeness of God and possessing an inherent dignity that demands respect and reverence. Human persons are treated differently than animals or things because of the awesome dignity they have been given by God, a dignity that is also accessible to human reason.

The starting premise of the book is that man is inherently open to life, a quality of human persons that is "built into us" and acts as a foundational argument for everything that follows. It builds on the author's previous works but consciously streamlines complex arguments to make them more readily accessible to a broader swath of readers. Adopting a deeply holistic and personalistic approach, the work embraces what the author rightly proposes as "a different kind of bioethics: a bioethics of the person." It is also highly "personal," in that it reflects the lived experiences and wisdom of the author himself, making the text readable, relatable, and even poignant.

As a good guide, Etheredge leads his readers along, inviting them to engage the very questions at the heart of every human life, and only then teasing out answers that satisfy both the mind and the soul. He does not rush to "teach" us what he has discovered but patiently walks us through the process by which such wisdom becomes our own personal discovery as well.

A lovely theological and spiritual thread runs through the whole of Etheredge's work, whereby the ethical conundrums posed by bioethics cannot be fully understood except in relationship to God's creative and redemptive plan for humanity, and indeed apart from his personal relationship with each of us. While the use of reason is essential for any serious bioethical enterprise, it is necessarily deficient and fatally incomplete — the author contends — if divorced from vital reference to God and his plan for the human person. This begins not with an evaluation of human actions and their moral dimension, but with the nature of the human person himself. We cannot comprehend who we are and who we are meant to be apart from our relationship with God. This theological dimension of the human person affects us as intelligent and free moral agents but also affects our understanding of who our brothers and sisters are, and how they should be treated as children of God. Moreover, at the author's skilled hand, theological abstractions — whether dogmatic, moral, or sacramental — are brought down to earth and made real and present through their application to our common human experiences.

None of this should suggest that *A Shorter Bioethics* constitutes "light reading." It challenges the mind and heart because of its philosophical, theological, and spiritual depth. It is a work not meant to be dispatched in an afternoon's read at the beach, but through a series of literary encounters taken slowly and intentionally, while savoring the author's gift of self through the written word.

And a true gift it is, for those blessed to receive it, and those blessed to pass it along.

Thomas D. Williams,
professor of theology at St. John's University
and author of *Who Is My Neighbor?*

Introduction

Francis is a Catholic married layman with eleven children, three of whom are in heaven. He has been writing investigations for many years now; and, owing to being driven from his last post, he has dedicated the last ten years to this task. Providence, therefore, provided an opportunity to write full time and seventeen of the eighteen books published have come together in this time.

Working alone is never alone, and he has collaborated with many people, from ordinary laymen and women, deacons, priests, bishops, and bioethical experts. Some have contributed to the books themselves, either a Foreword, an Endword, an Introduction to a Chapter, written a poem, or endorsed or reviewed various books. Moreover, even when we work as an independent writer of non-fiction, there are many who contribute via their articles, books, criticisms, including publishers, interviewers, and indeed via that advice he receives he is enabled to approach his subjects in a simpler format. Thus, over the various books there is a colossal amount of available material which, therefore, cannot be reproduced here; and, that being the case, I apologise if anyone feels that their contribution has not been acknowledged. However, the books themselves are replete with thousands upon thousands of references; and, of course, work is constantly being generated. At the very least, then, I hope these essays stimulate the search for reliable, foundational arguments, which he hopes will outlive the toing and froing of different ideas and arguments and give the generations to come firm ground to build upon. For, in as much as some think that reality is like sand, shifting in such a way that there is no constant to human nature, the stability of what exists speaks for itself and invites us, constantly, to visit and investigate what exists.

However, it is clear that many of these investigations are far too long for most people to read. Therefore, "A Shorter Bioethics" is an attempt to take the main arguments, without footnotes, and to set them down as concisely as possible. Furthermore, the structure of the book, as you will see from the "Contents", takes a beginning from being open to life as this is, as it were, built into us and acts as a foundational argument for what follows.

He is very grateful, too, to the two proofreaders who have volunteered their services, free, over many years. One of them, Martin, has now gone to *The Father*. But there are many who do not make the headlines or the footnotes but who, in one way or another, contribute, stimulate, and help. Thus, beginning with my wife and children, and then going on to others, whether a part of his Church-Community, it is good to credit many people for their help over many years.

These are the principal texts from which this set of arguments come. These are, for the most part, books; however, in one or two instances there are relevant essays, some of which are not cited here.

Introduction

Why am I a writer?

Why am I a writer? If you questioned a bird about its wings. If you questioned an animal about its legs. If you questioned a fish about swimming, what would it say?

As a youth, I fantasized about being a footballer and played for a village team, worked as a farm worker, drew and tried to be an artist but fell at the first hurdle, after a foundation year at art school. In response to the question and supposition, why did I want to be an artist, was it because my father was one? I left the interview in confusion and only ever drew and painted intermittently again. I started reading more than comics, which I had been doing up until sixteen, and started reading about learning to be a surgeon on a battlefield. Then came the questions, in my lonely bedsit, at 16, as I tried to play tennis against a wall and work in an office, and bought some paints and never used them. Eventually I returned home, yet again, getting nowhere with who I was and what I wanted to do and so, suddenly, I started writing about a kind of psychological stream of consciousness.

I started writing before word-processing and it used up a lot of paper, glue, cutting and literally pasting, crossing out, arrows, extra pages and the interminable impossibility of ever getting further than a few more extra pages and the end never in sight. After being totally distraught about editing, finishing and accomplishing anything in writing, I abandoned the whole, by now bulky projects and threw them away, giving one of the shorter pieces to a complete stranger and never seeing it again.

Then there were years of travelling, of trying to be a journey man, maintenance man, monk, priest, furniture maker, sculptor, nurse, helper to young people and, in the end, abandoning it all and returning to study design but then abandoning that and, finally, starting to do philosophy, law, psychology, and theology and starting to write again.

Many years later, when driven out of work, having had one book published, and a few articles, I began to write regularly, listening to criticism, remembering an old tutor saying a book is a collection of essays, and finding a publisher and then, again, another publisher and then more books.

Why am I a writer? Writing is a way of investigating the world, whether internally, taking up questions about identity, questions which had bitten deeply into my heart, like discovering that a woman had destroyed my child because she thought it was a "blob of cells", and searching for the truths that add up, like bits of the true cross, to the mystery of sin, suffering and redemption from Christ. And, one day, the Lord took pity on me and said, through the Catechism of the Catholic Church, If God can make all that exists out of nothing, He can make a new beginning for the sinner and that sinner me. And he did.

Three kinds of published work

The following titles are now on Amazon, in production or, as with this book, in the process of being written; but I list them here in three groups: Scripture; theology and philosophy; Bioethics; and autobiographical; indeed, the third group began

Introduction

with a prayer in front of the Lady Chapel at our local Catholic Church, in the hope of writing a more widely accessible read.

Scripture: A Unique Word (2014); A trilogy on Faith and Reason called, From Truth and truth: Volume I-Faithful Reason (2016), Volume II-Faith and Reason in Dialogue (2016), Volume III-Faith is Married Reason (2016).

The Human Person: A Bioethical Word (2017); Conception: An Icon of the Beginning (2019); Mary and Bioethics: An Exploration (2020); Reaching for the Resurrection: A Pastoral Bioethics (2020); and Unfolding a Post-Roe World (2022); and Human Nature: Moral Norm (2023). Furthermore, the *National Catholic Bioethics Centre* of America has invited a contribution to a second edition of a book on *Human Embryo Adoption* (2025); my contribution is called "The Annunciation and Embryo Adoption" and argues that embryo adoption is a Marian participation in the saving work of Jesus Christ. The Zygote of Christ and the Mystery of Man is co-authored by Francis Etheredge and Dr. Elizabeth Rex (2025). The following books, while they have autobiographical elements, are principally about the impact of the times on our youth. The Word in Your Heart: Mary, Youth, and Mental Health (2024); Transgenderism: A Question of Identity (2025), followed by Transgenderism Two: Evidence and Identity (2025). A final work in this section is a short, summary book, taking a fresh look at the previous body of work; it is called A Shorter Bioethics (2025).

The Family on Pilgrimage: God Leads Through Dead Ends (2018); The Prayerful Kiss: A Collection of Poetry and Prose (2019); Honest Rust and Gold: A Second Collection of Prose and Poetry (2020); Within Reach of You: A Book of Prose and

Prayers (2021); and now Lord, Do You Mean Me? A father-Catechist!" (2023), An Unlikely Gardener: A Book of Prose and Poetry (2023).

Finally, I wish to thank Mr. Martin Higgins, MA, now deceased and, more recently, Dr. Anthony Williams, for all their proof reading. Furthermore, for all those who have contributed reviews, endorsement, or text contributions, and there are too many to mention by name! Finally, thank you to Dr. Sebastian Mahfood who has taken on the work of publishing the majority of these books.

All that is published to date can be found through this webpage:

https://enroutebooksandmedia.com/francisetheredge/

Preamble

What is Bioethics?

Our understanding of the human person can be very fragmented, reflecting the multitude of disciplines through which we study nature, whether globally, according to the needs of particular plants, or animals. Similarly, it is possible to be "act-orientated" when it comes to bioethics: to examine particular crises, whether it is the beginning of life, suppressing life, whether contraception, abortion, or suicide, whether assisted or not, or any number of other "bioethical actions". In other words, we scarcely look at the whole, as if the whole is out of reach, or is so mechanical in understanding that there is almost nothing to see beyond mechanisms, money, or the chemicalization of the body.

But the more it is possible to reflect on these questions from the point of view of the human person who either acts, or is acted upon, the more it is possible to envision a different kind of bioethics: a bioethics of the person. In other words, the human person is a bioethical word. The whole person is both immersed, as it were, in being wholly human and, at the same time, wholly expressed in being a wholly male or female bodily human being-in-relationship. An early version of this kind of understanding was the concept of the human person being a psychosomatic whole: an integrally unified, ensouled body, or a bodily expressed soul.

Listening and being listened to

One of the most poignant expressions of human identity is being listened to: that we articulate who we are in relation to another or others; and, indeed, when our capacity to communicate, closely, with those close to us, collapses, then we are approaching death by dialogical starvation. Whether we are a "stay-in", a retired workman who never goes out, a soldier or builder away from home, if the communication is not kept lively and open, there is a possibility that we will suffer a kind of social withering-freezing. As we are human beings-in-relationship, it is as if the oxygen of life is leaking out and is being slowly replaced by non-smelling, but poisonous gases, so that almost unnoticed we slip into a kind of body-bag, going from unwilling to unable to go out and risk rejection anymore.

So the very relationships that are denied through taking contraceptives, having abortions or any number of fake identities, indicate that we are suffocating – not just through losing contact with others but through losing contact with others, we lose contact with ourselves.

The questions which really come to the fore are what drives anorexia, a slimming disease, changing our facial appearance, in a world obsessed with how we look? What questions of meaning are at work when a person struggles to understand the loss of a child, why we run through relationships as if it is a menu, when all the while we live in a culture where hooking and unhooking can be overnight or over the discovery that there is nothing more to talk about? And so when the abandonment comes, as our children are no longer in contact with us, or we are estranged from our parents, or our efforts to find fulfilling work are so futile as to be, if not funny, frustrating to the point of despair, we may well be untrained in the question of what does it mean: What is the point of our suffering?

Suffering and the search for meaning

In a certain sense, human suffering has a point. The point of human suffering is that it is like the surgeon's scalpel, it cuts away at our human effort, the masking of our problems, exposing our hidden addictions, until the pain becomes visible, if not unbearable, and we ask is there a meaning to it? If we do not engage with the meaning of suffering it will suffocate us? If we remain at the level of human effort, of bearing with it, of trying to solve it for ourselves, of changing diets, going to the doctor, or even planning an end to it, then we have not reached the point of puncture. The puncturing of our efforts is when our hearts start opening onto a different horizon: the horizon of the depths of our being human and in relationships which, perhaps, we have neglected all our lives: the relationships which arise out of our very existence: of standing forth from non-existence – of not being the cause of our own existence.

Suffering, then, is what strips away our psychological strategies, our claims to be able to solve our problems, our reliance, ultimately, on ourselves. But this stripping is not pointless, just as the surgeon cuts to heal, so does suffering cut to drain the puss of resentments, unforgiveness, and being unforgiving. Suffering slowly and surely puts us in front of the possibility of death, of whether or not it is a precipice over which we can fall, of whether or not it is possible to live life in a new way - but not because of who we are or even because of those around us. Rather, suffering stretches our hearts so that we want to break out of its shell-like hardness and calls to us to cry out. Is there anyone there?! Is there a God?! Why am I in this wretched mess?!

But above and beyond the differential diagnosis that tells us whether there is any cancer, new illness, or question that cannot be answered by a doctor except in terms of pills, there is the word of God. The word of God is, as St. Jerome said, a word which leads to Christ: Ignorance of Scripture is ignorance of Christ; and, I might add, ignorance of Christ is ignorance of ourselves. In other words, there comes a time when we need what will make us transparent to ourselves. We need a lemon, as it were, which will reveal what is hidden in our hearts. We need a dialogue with one who knows us better than we know ourselves.

The Christian depths of human being

Bioethics, then, brings us to the sensitive point of pain and, indeed, entails the realization that communication is about admitting what we are actually feeling and thinking. At the same time, there are elements of human identity which go deeper and involve perceptions of ourselves which bring out, as it were, "the Christian depths" of the human person. I can look back, now, after about fifty-six or fifty-seven years, and see the child I was at thirteen or fourteen and "hear" the swearing in my heart that I would not admit the pain of being humiliatingly caned in front of the class, although sometimes in an office, because of failing the class tests and being bottom of the class. In other words, there was a spontaneous reaction in my heart, even without thinking it through, that I would suppress the pain of being caned. This suppression of emotion led not just to the suppression of the pain, but that of the memories and emotions too, such that it was not until my later adolescence that I began to wonder what was wrong with me that I did not remember what had happened in my childhood.

Thus, the deeper we go into the full, wholesomely human nature of the person, the more we discover that the depths of our sufferings open up dimensions of the human heart that are "outside" the brief of the medicalization of suffering. Indeed, if all the medicalization of suffering entails is the use of drug-type suppressants of consciousness then it is totally inadequate to address what is of a different order of problems. Indeed, it is like trying to use a virtual wind to blow out a candle: the two entities are too different to interact; and, in fact, the preoccupation with trying to make a virtual wind blow out the candle, is like working on a screen while the house burns down.

The choice, then, to be chosen, is to go further into the announcement that there are spiritual depths to the human person or to reject the spiritual expression of the human person. Clearly, however, as experience often confirms, it is impossible to go beyond what human effort can accomplish if we do not depend on what is more powerful than human effort, namely God. In my own experience, then, of suffering repeated failure when it came to forming lasting relationships, particularly with a view to marriage, it became obvious that there were three possibilities in my life: the deterioration of my mental health; suicide; and going from wrong to wrong which, actually, is from sin to sin. Usually, then, the call to faith arises out of the preaching of the Gospel of Christ (cf. Romans 10: 17).

I am convinced, however, that as a fruit of other people's prayers, the Lord spoke to me through the *Catechism of the Catholic Church* and said: "That if God can create the universe out of nothing – He can make a new beginning for the sinner" (298). And, without realizing the power of this word until I returned to an adult formation in the Catholic Faith, called the Neocatechumenal Way, and discovered that I was no longer afraid of the sufferings of marriage, that had kept me from "entering" the lobster-pot of no-return and, within a year, I was both married and have now been married for nearly thirty years with ten children, two of which are in heaven with a third one of mine.

Our sufferings, then, cannot be answered by an inadequate account of the human person; and, if what we know or think does not answer our questions, then it is time to rethink what we do think and, possibly, to ask God for the help He alone can give. And, while it may take many years to unravel, heal, persevere in the help the Lord God gives us, our whole mentality begins to reflect the good He is doing in our lives. Moreover, as with my wife and I, we are part of a community with whom we share the development of our Christian life and belief. Thus, little by little, the impossible becomes possible and, instead of rejecting the possibility of driving my children about, I begin to see that this time together is also a real help to the parent-child relationship. And, what is more, that this relationship to our own religious faith begins to become a "gift" for the other, whether that "other" is our children or other people, or both!

Chapter One

Faith and Reason

We live in the long shadow of eternity, as it stretches back from the future of mankind as well as in the light from the dawn of creation; and, indeed, it is as necessary to live in view of my own death as it is to draw on what can be done in each day of life.

When it comes, then, to a variety of investigations, at the very least there is a prayer to the Author of life who, having created all that exists, knows it more intimately than ever we will. It is not surprising, then, that a prayer is often a part of the preparation and the action of study, research, and writing. Moreover, given that I am called to live what I do, it is no surprise, then, to pray daily to the *Lord of Life* to inspire my turning away from sin and my turning towards Him who is *The Way, the Truth and the Life* (Jn 14: 6). Furthermore, this lived wisdom will be called upon, whether it is our youth with mental health problems, transgenderism, or the questions of when does life begin, what is morality and, indeed, what is a human being?

At the same time, however, conversion is a turning towards what exists, what "is" and, while our whole lives are a turning away from falsehood and a turning towards truth, there is nevertheless a constant reference point: What is the reality of the human person? It is a fiction, in the end, to suppose that we can investigate the whole reality of the human person and not include what constitutes that whole: conscience, reason, perception, evidence, experience, our emotional life, decisions, judgment, prayer, truthfulness, love, and the help of others: of sharing ourselves in that dialogical exchange of gifts out of which we live. So, the more we reduce what exists to "matter", the easier it is to manipulate it but the further away we are from the reality in need of investigation. Is it unsurprising, then, that even intelligent and well-trained people, doctors, clinicians, professors, teachers, and researchers are, in effect, turning away from reality in the search of what conforms to their predetermined "idea" of reality rather than examine what exists?

Change and Constancy

One of the ancient questions is what kind of being, what kind of object, what kind of existence is there to "reality"? Is it all flux and change, like water running through time, always different? But even water running through time has a pathway, as it were, a riverbed, so that even if what flows through constantly changes, there is nevertheless a constancy to the riverbed: to what makes continuous change possible. In other words, even constant change has what constitutes a framework within which we think and work. So, while the image of water works well enough for change, yet change runs through us in the form of consciousness, and human consciousness would not exist if there was not the constancy of the body's expression of it. Similarly, the "infinity scroll" on a phone, being able to roll from image or text to image or text indefinitely, depends on the existence, connectivity, and energized structure that makes possible an ongoing supply of inputs. The phone itself, then, is a

Chapter One: Faith and Reason

definite object, like ourselves, capable of an incredible amount of what passes through without a lot of detailed reflection, so much as we see, feel, think, judge, act constantly, whether superficially or at a slower, deeper pace. And, if we trash the vehicle, either ourselves or our phone, then we lose the capacity to carry and to communicate this stream of intentional consciousness: a consciousness that is always seeking to listen, to know, and to communicate the truth in love as wholesomely as we can.

What this brings us to is the whole universe. For it is the whole universe that makes possible both the constancy and the changes in what exists. At the same time, the change that comes to exist with the coming to exist of the human person, is a radical change which, according to the original meaning of the word, takes us back to a new beginning: human conception. But not just to a new beginning, but to a beginning that has had to have an act of creation at the root of it, namely that the human person originates from an act of God at the beginning of each one of us. But this is not fictional biology, in that except at the beginning of the human race, each one of us has a biological prehistory, recognized as rooting us in the human race and in the specific relationships, first and foremost to our own parents and family.

There is a claim, which many make, that there is a process of underlying change which is so slow that it has to be calculated in timescales in which it is impossible to detect specific instances of change, except by the conjecture that one kind of creature was the forerunner of a different kind of creature. In one sense, this is an irrefutable argument, in that we cannot observe, directly, distinct and cumulative changes that constitute one kind of animal becoming another kind of animal. At the same time, if there is no recognition of the difference between the existence of what makes thinking possible, judging between options, loving another person then, simplistically, there is no obstacle to the claim that there is an uninterrupted development from matter in its early state and a human person. That simplicity, however, disregards the integral whole of the human person who is, in effect, a personal complexity which cannot arise out of what is simpler than his or her own existence without a cause that goes beyond both.

The very existence of agriculture, craft, literature, science, medicine, philosophy, religion, hate and love, all testify to that identity and depth of the human person which is not "visible" and "evidenced" in any other type of enfleshed being. Thus, while there is plenty of evidence of intelligence, in all creatures, proportionate to their complexity and size, whether it is finding a mate, making a nest or other domestic habitat, recognizing food, rearing young, there is no comparable flexibility to that of human creativity.

The development of cooking, for instance, has such a subtlety about it that it is, and once was called, domestic science. As this recognizes not only the creativity of taste and nutrition, but vast and varied combinations of what, together, makes a meal or a desert; and, coupled with temperature, implies a vast knowledge of recipes, with what does not need to be cooked, and the wholesale benefit of all this to human and animal health. So, even if modern science has succeeded in many areas of life, it is possible to regard cooking as one of the most ancient of sciences. Indeed, given the connection between health and food, cooking can almost be called the "cradle" of the sciences as, indeed, a lot of implements would have been inspired by the very development of that science. Clearly, however, what is used for cooking, hunting and for conflict, overlap; and, as always, it seems as if the misuse of knowledge accelerates with the accumulation of it. While what is understood and expressed in recipes, grows with the generations who received them and who transmit their own additional variations, the confluence of the ingredients and dishes of different cultures confirms and multiplies the amazing grasp and versatility of the human senses and the penetrating intelligence of the people who continue to develop it.

The Contribution of Faith

At the very origin of human existence there are questions. How can what is materially biological give rise to what actually goes beyond it? Indeed, how can what is materially mineral or chemical give rise to what is materially organic and living? In other words, the very existence of the human person, taking up as he or she does, the very constituents of the universe, is a marvel in itself. Similarly, the capacity of what is materially organic to take up what constitutes the ingredients for growth, like the seed takes up the nutrients in the soil, is a marvel to behold. Similarly, the very capacity of what is capable of movement, not just the movement of what is rooted in the ground to the sun, bending with the wind, or possessing more developed sensory responses like the venus fly-trap, is itself a challenge to our understanding. In other words, the animal type of creature, whether bacteria, bugs and slugs, or the bigger and more mobile creatures like the lion, the tiger, the elephant and the eagle, are not only capable of utilizing the mineral and organic vegetable or animal resources of the earth but transform it into the characteristics of their own growth, reproduction and, where it applies, the care of their young. So these great steps pose their own questions. What kind of principle of organization makes these different kinds of creatures, materials, or existing objects whole, unique, but capable of all kinds of cooperative acts, combinations and activities? But, for the purposes of brevity, what is of concern here is the way that the existence of a human person speaks of a different kind of origin to that of an animal, a plant or the materiality of what exists – except that they all have a common question which is, in reality, what can cause each kind of being to exist?

So, in focusing on the human person, there is the ever-recurring question of why there is something and not nothing. Indeed, why there is a unique, personal being, who comes to be named Cain, or Abel, or Mary, or Elizabeth, each of whom is either male or female and contributes to the dialogue within and between the sexes. So the question of human identity is distinguished from the identity of what otherwise exists. The human person manifests both an identity that is a whole, namely, that of being a man or a woman and, at the same time, that identity is uniquely individual, not a "rote" repetition of being human. At the same time, the individual is ordered to relationship, having come to exist from relationship, the human person comes to exist in and for relationship.

Chapter One: Faith and Reason

Focusing on the human person, then, focuses us on the nature of human relationships, which raises the question of what constitutes human relationships? In other words, why are we not indifferent individuals? Why is loneliness, widowhood, the desire to marry, the vocation to the religious life, the belonging to different kinds of groups, beginning with the family, so characteristic and common to human beings? Thus there is a kind of transcendent identity to the human person: that the human person goes beyond a functional relationship to one another, although relationships entail collaboration, whether in marriage, family life or the workplace or as members of the human race expressing a common relationship to each other or to an "Other". So faith, taking the very human ground of our existence, shows the personal act which constitutes the very act of personal existence, to be the act of a Personal God: A being whose very identity makes possible the likeness of human beings to the "Other" who brought each human person to exist.

But faith, the Christian Faith, is more than knowing that we exist from a personal act, of both our parents and "One" who can cause to exist what does not exist; it is the "in-breaking" in human consciousness of the "Person" who brings to exists, what is personal, by bringing to exist a personal relationship – not just between each of us and what is known as God but also each of us and each other as we all have in common a personal cause of our existence. The Christian Faith, then, takes up all that is characteristic of God in the history of salvation and, at the same time, shows that that gratuitous love, that overflowing, abundant love, of which we are a fruit, is re-expressed in the person of Jesus Christ who "saves" and "restores" who was lost and injured by the actions that deface the humanity to which we are called.

Faith, ultimately, comes as a gift to complete the intelligibility of relationship, out of which we came to exist, and within which we struggle to remain, and for which we are called to respond to "Who" loves us. But faith overspills, like the very act of all that has been brought to exist, comes to exist, so that we see more and more the loving intelligibility of all that God has done, is doing, and is bringing about. All of which is expressed in the goal of an eternally reciprocal love "like" that of God's own reciprocal love between God the Father, God the Son, and the God the Holy Spirit: not three gods but One God in-relationship.

That unity, then, of faith and reason, of reason and faith, is as reciprocal as the very mystery of all that exists is both intelligible and the expression of relationships which need to be revealed beyond what we are capable of discovering. So we can see that a disorder exists in our very being, and in the very being of all that exists, so what exists communicates the mystery of a cause entering human history which corrupts the original gift of the relationship between God and His creatures. And, at the same time, we can see that there is a work of God which goes beyond reclaiming what was lost, which goes beyond restoring what was disordered, to what is hitherto only indicated and intimated in the glory of a final communion between God and man, male and female, and all His creation. Thus, whatever we do, with whomever we do it, is all with a view to a final consummation of communion between each of us and all of us and God.

Chapter One: Faith and Reason

Ethics-in-Relationships

In view, then, of the whole human nature that each one of us possesses, there is a challenge to communicate that wholeness simply, not drawing on what appear to be, but are not, two complementary parts of the whole, namely, body and soul. These are not complementary as man and woman are complementary, that is, man and woman are two different subjects, are two different, but complementary whole persons. Whereas body and soul are like the interior and the exterior of the whole person. God creates the soul the instant that the body comes to exist, such that there is no instant of time that the body existed, before it was ensouled. In other words, the whole human being comes into existence in the one instant of the whole coming to exist. Thus, while the one is interior to the other's exteriority, the exteriority of the body communicates the interiority of the soul. This is true, theologically, in that if Jesus Christ's body was created a moment before His ensoulment, then given that a body needs an animating principle for it to be alive, then the body of Christ would be alive with less than a human soul, which is both beneath the dignity of God and, at the same time, indicates Christ's soul supplanting a less than human agent of animation. To suppose that God would "invent" a relationship to maintain the body of Jesus Christ before His ensoulment is not only to suppose that God's original creation did not take account of the possibility of the Incarnation, but is also to suppose that there is a distinction, other than Christ lacking the contribution of a human father, between His conception and that of the rest of us, post Adam and Eve. Similarly with Mary, and this applies to Christ's conception too, if grace requires a human subject to be the beneficiary of it, then Mary was conceived without original sin the moment she, whole and entire, came to exist. Otherwise, Mary, like Christ, would have been united to matter, whether biological or not, that was without the perfecting grace that accompanies the presence of the soul and permeates the whole body.

So, how does all this bear on "ethics-in-relationships"? Ethics, expressing the moral compass of the human being, is an expression of the whole person-in-relationship. Whether that relationship is to the environment, other creatures, ourselves, other human beings, or God, these relationships are wherein our action is expressed; and, in being expressed, either does good and avoids evil, or does not do good and does do evil. In the case of Jesus and Mary, the very integral nature of their graced humanity determines, in other words, that the very good of their whole being is expressed in a perfection of their moral behaviour. That, in reality, their relationships are lived out of love, which is the proportionate response to the good of the human person. Thus these actions of Jesus and Mary express truth, the right judgment of what is in a person's heart and how to respond to them, which is the perfection of human reality in motion: a perfection expressed from the very heart of being perfectly human. Morality, the perfection of human action in which Christ's injunction is fulfilled, which is that *without me you cannot do anything* (John 15: 5). In other words, unless Christ's grace acts in us, we cannot do the good that befits salvation. This is why, however imperfectly, the Christian is called to immerse him or herself in the word of God and prayer to the

Chapter One: Faith and Reason

extent that then the person acts out of Christ's love for the sinner, both the sinner who acts and the one whom we seek to help.

In sum, the goodness of creation expresses the Goodness of the Creator as indeed does the love of Jesus Christ and the action the Holy Spirit. So it is with us: our deeds flow from what is in our hearts. As Christ said: 'It is what comes out of someone that makes that person unclean. For it is from within, from the heart, that evil intentions emerge: fornication, theft, murder, adultery, avarice, malice, deceit, indecency, envy, slander, pride, folly. All these evil things come from within and make a person unclean' (Mark 7: 21-23). Conversely, if the heart is occupied by God, as He makes His home in us (cf. John 14: 23), so our deeds express more of the perfection of God than of our own, imperfect nature.

On the one hand, then, we are called to use reason and evidence for what is good, true, and expressive of love; but, on the other hand, we are to invite the Lord to perfect our hearts, that what arises is good fruit (cf. Matthew 7: 17). In what follows, there is the possibility of discussing, further, the human action of procreation and its perfection by the help of God.

Chapter Two

The Whole Human Being and Contraception

As the investigation, as it were, of human anthropology, of what makes a human being a personal being from conception unfolds, we are in a position to consider the unity of the human person more and more fully. Clearly, we live in times when the unity of the human person is completely undermined, and destructively, to the point that the body is treated as a kind of "external" shape with no relationship to the internal constituents of the person. Neither is the human body understood to be a psychological expression of a personalized, human sexual identity, which is expressed according as the person is male or female. Unless identity has a root in what exists, then I am what I dress up to be, which can be anything from a monkey to a spaceship made of cardboard. Or, as some claim, I am what my performance says I am, which could be interpreted to mean if I dress as a doctor, carry a stethoscope and talk about diagnoses I am, without qualifications to corroborate it, a doctor.

The Sacrament of Marriage

What is to be discussed, here, however, is precisely the relationship between the root and fruit of being a man or a woman; and, indeed, their complementarity, especially in view of marriage. The Catholic *Sacrament of Marriage* is an outward sign of the inward love of God that unites, unbreakably, the man and the woman. As Christ said, 'What God has joined together let not man put asunder' (Matthew 19: 6). Let us note, "What God has joined together"; and, therefore, what takes an act of God to unify, is not just a union natural to man and woman; but, rather, it is both a natural and a supernatural union. Or, better, a union which is both natural and supernatural. For, if it takes an 'act of God' to form the 'one flesh' of man and woman in marriage, it is because the spiritual union, formed in the marriage of two baptized Christians, is a union which both goes beyond what is natural and communicates what is beyond the reach of our humanity while, nevertheless, being ordered to it, namely that Christian marriage expresses the mystery of the union between Christ and His Church (cf. Ephesians, 5: 28-32). The union, then, between husband and wife is a communion which expresses, in the dialogue of their salvation, the relationship between Christ and the soul of the Church, the Holy Spirit. In other words, just as the communion of the three persons of the Blessed Trinity gives rise to creation, so the communion between husband and wife gives rise to a service to life.

So, Christian marriage, expresses the mystery of the marriage of Christ and His Church; the mystery of Christ and His Church, is a mysterious union, fruitful in the gift of making us children of God. Thus Christian marriage is both a covenant and the family is a *domestic Church*. Christian marriage is a covenant, like the covenants of old, expressing a relationship to God, His presence, and a promise of His faithfulness. Christian marriage is the foundation of the *domestic Church*, the expression of the universal Catholic Church in the particular reality of marriage and family life. Christian marriage is fruitful, too, in cooperating with the mission of the Church to give glory to God, through witnessing to the providential action of God in their lives and to make disciples of Jesus Christ. The disciples of Christ can come through the possibility of conceiving and bringing forth children, who are called to grow up to be, through education and a training in the Christian Faith, living members of the people of God. Or by serving life through adopting, for whatever reason, children who need parental help. Or, again, by any service to life for the good of others and the Church. Or, indeed, some combination of any or all of these services to life.

The nature of marriage, at once like the nature of the Church, is composed of both a natural and a supernatural complementarity; and, as with the Church, which is the one *Body of Christ,* so are husband wife one flesh. Similarly, if the *Body of Christ* is not two things, but one, so are husband and wife not two people, but one flesh; and, being one flesh, they pray and act together. In a certain way, although for tragic reasons, our contemporary culture throws into relief, more and more, the mysterious dignity of husband and wife. Indeed, given the claim of some that the woman and the man are to be compared to a bull and many cattle, the Christian equality of husband and wife stands a lonely watch on the gift of woman. Similarly, given the atrocious exploitation, or even sex-ploitation of women and children, how much more is the equal dignity of woman expressed in the amazing, reciprocal self-gift, of one man one woman. And, what is more, that this human dignity is, as it were, a reflection of the divine equality of person to person in the mystery of the blessed Trinity. Similarly, the child being born of the sacrament of the marriage, preserves the inviolable dignity of the child. Just, then, as one man and one woman communicate an irrevocable equality, so does the child born of marriage.

Albeit, in our fallen world, all are fallen and need the help of God. But, even so, it is very striking that marriage and family life expresses, in its own way, the unique dignity of each member of it. Furthermore, adoption and the service of life seek to restore, wherever possible, the wounding of this dignity.

Being Open to Life: a practical spirituality

Thus, being open to life, while arising out of the dialogue and prayer of marriage, develops out of the reciprocal knowledge of the spouses, their coming together. According to the *Book of Tobit*, Sarah and Tobias prayed together as husband and wife (8: 4-8) and, similarly, this can inspire married couples to pray before coming together as husband and wife. On the one hand, then, there are many considerations to take account of when it comes to being open to the possibility of children, some of which are the health of husband and wife, the practical circumstances of marriage, whether living in one room or a flat, on a low income, and the prayer that opens the couple to the providence of God. Each of our own children were prayed into existence, sometimes after extensive discussion and consideration of our circumstance but also, taking account of how many times God had helped us in our lives with money, illness, accommodation, and work. On the other hand, being open to life is not just being aware of the possibility that any particular act of marriage will bring about the existence of a child; rather, it is that awareness but in the context of a spirituality of service in the light of the providence of God.

What I want to focus on, now, however, is not just the possibility of conceiving a child out of the dynamic communion of husband and wife, but the marital unity out of which this arises. A particular consideration is, then, both the cooperation of the husband and wife in the service of life; and, indeed, this is where an anthropology of gift, an account of the gift of personhood, breaks through the mechanistic understanding that conceiving a child is about "imposing" a biological reality on the couple. Some people criticize the Catholic Church for suggesting that the inseparable connection between marital union and procreation is an imposition of a biological reality, the nature of the woman's cycle, as if it is a moral principle, on the married couple. But what if the inseparable connection between marital union and procreation is to communicate gift from gift: that the gift of the child is a gift from the reciprocal gift of husband and wife. Just, then, as there is a choice in entering marriage, as coercion or deceit can invalidate it, such that the love of husband and wife is a love that seeks to be reciprocally given and received by each of them, so a child is a gift not a right. What underlines the nature of the person-as-gift, is the natural uncertainty as regards the conception of a child. In other words, just because husband and wife are open to the possibility of conceiving a child, there is no certainty that a child will be conceived.

What we have, then, in a marriage, is a personal expression of the interpersonal love of the Blessed Trinity, out of which God freely created the universe, all that is in it, and made man, male and female, in His own image and likeness. And let us not overlook the biblical, grammatically gendered language, which both gives God as a plural, with three distinct names and, at the same time, the Hebrew writer used heaven, as a masculine noun, and earth as a feminine noun, to introduce the whole procession of creation, like from like, ending with man and woman. In other words, an interpersonal love of the Blessed Trinity, through which the interpersonal love of husband and wife, co-creates with God a child-person-gift. Thus each human being is an end in him or herself: a person-gift to enter into the interpersonal relationship through which he or she came to exist, both beginning in this life in terms of the child's parents and family and in terms of a relationship to God who brought the child to exist.

The cycle of fertility or infertility: An abstract data or an intimately integral source of self-knowledge for the woman

What, then, to step back a few paces, is the connection between all this and the natural cycle of fertility and infertility? On the one hand, there is the mentality that a woman's cycle can be separated from the woman; and, indeed, that is a part of the mentality of contraception: that the medicalization of a normal "part" of the woman is addressed as if it is a disease of some kind. And, if the menstrual cycle can be separated from the woman then it can be separated from the woman-in-her relationships – particularly in her marriage. But, to the contrary, the woman is an integral whole; and, in entering marriage, the woman enters fully, like the man, in making a self-gift of herself as he makes a self-gift of himself. In other words, neither of them, in full knowledge of the possibility of becoming a mother and a father through each other, withholds any part of their self-gift; rather, if they abstain from the marital act, they do so together and for a specifically grave reason and, if possible, only temporarily.

With respect, then, to the mentality that the biological rhythm of the fertility cycle is almost "extraneous" to the woman, the first consideration is that a woman's recognition of the reality of her cycle, whether it is "visible" and "clear" is a general indicator of her health. In other words, from the point of view of a woman's health, there is an enormous benefit to not taking anything which obscures this "health indicator". Or, if the cycle is not regular or there is another problem, then these symptoms are a reason to investigate. Whereas, putting it broadly, if taking an unnecessary set of hormone tablets obscures the natural indicator of a woman's health, not only is she deprived of this but the woman does not learn, either, to recognize her own cycle of when she is either fertile or infertile. And, to put it simply, the whole study, as it were, of what occurs, when, and with what, if any, complications or variations.is a foundational awareness that can only benefit the woman herself. Moreover, by contrast to millions, if not billions, spent on unnecessary drugs that pollute the waters of the earth, a woman studying with a woman is relatively cheap or even free. In other words, even in terms of the woman herself, the menstrual cycle is integrally expressive of her womanhood; it is not an added extra. What is more, the cocktail of drugs that can come with contraceptives, some addressing so-called side-effects, impacts the whole person and can indeed induce a certain kind of dullness, impeding the whole marital relationship.

More generally, on the one hand, the whole, integrated, functioning biological expression of the person passes threshold after threshold, showing forth the intimately present psychological processes. These bio-psychological processes, being latent and dependent on prior developments, become clearer and clearer as movement, expression, and responsiveness develop, even in the womb, but more and more obviously as the child grows into an adult. Thus the integration of the bodily expression of the person, characteristic of marriage, is a magnificent outcome of the more general unfolding of human identity. Therefore, the receptivity of the woman-as-wife, entails contributing the whole of her identity in marriage, just as the activity of the man-as-husband contributes the whole of his identity in marriage. Between them, then, they develop a sensitivity to that awareness which is about coming together as husband and wife; and, indeed, it is clear to them that this enhanced perception of each other is a gift to them both. In other words, the reciprocal gift of husband and wife arises out of the psychological interiority which is inscribed in the very structure of human sexuality. We are not, therefore, examining an expression of the sexuality, of either the husband or the wife, as if it is external to the person; rather, the intimate nature of marital communion indicates an interiority to human sexuality whereby it expresses the personhood of both husband and wife.

While, at the same time, there is a necessary call to prayer and abstinence, if the family size and circumstances warrant it, the married couple are nevertheless in possession of a certain golden intimacy. They understand the original gift of man and woman, husband and wife, and are free in the reciprocal gift of marriage. However, if a difficulty does arise, they are prepared for a period of conversation, abstinence, and prayer, precisely because this is the familiar ground out of which marriage begins and unfolds.

By contrast, a culture that presupposes the availability of either husband or wife, is a culture that does not understand human illness, sin, and suffering. For, it is just as unrealistic to suppose that there will never be times of abstinence in a marriage, as it is to suppose that there will not be unemployment, crises of health, housing, problems with children, or even infertility, temporary or life-long. In other words, a mentality that supposes a pill can answer the reality of life, is as unrealistic as it is destructive of personal responsibility; and, therefore, is more likely a cause of divorce, infidelity, and the abandoning of marriage than is probably realised.

Chapter Three

An Actual Beginning: Conception

What do you suppose constitutes a beginning? The moment a pen touches the paper. The moment a match is struck. The moment a spark ignites a gas. In a certain way, the beginning of life draws on these different moments, whereby contact between two ingredients, as it were, initiates a change. Whether, then, there is a child conceived through the marital act, twins emerge, a woman's egg is fertilized by a man's sperm in a glass dish, there is the beginning of a human, embryonic child, and all the relationships which come into existence too, which express the integration of the child in the immediate and whole human family.

It is possible, however, for the claim to be made that there were prior moments, of the thoughts that had formed and, like a chemical reaction, spilled onto the page. Or the striking of the match, never mind it being taken from the box. Or what caused the spark, on the one hand, or where the gas came from on the other hand. However, all these additional notes, do not give us the moment of contact; and, as such, it is the moment of contact that constitutes a beginning. By contrast, some claim that there is no moment of beginning and attempt to assert that there were several; for example, the pen had to be made, as did the paper, as did the preamble that led to wanting to bring pen and paper together. However, all these prior moments, while they are valid and express the truth that each item in the sequence had to have a beginning, they do not constitute the beginning of the contact between the paper and the pen, the striking of the match, or the igniting of the gas. In other words, it is possible to take what we know and to throw out a kind of disguised version of it, claiming that we do not know when the person is conceived.

However, this is ingenuous, because the origin of the sperm, the ovulation of the egg, are not the beginning that their coming together constitutes. Therefore, let us not confuse preliminary causes with the causation which arises from the contact of the sperm and the ovum or egg. But lest we think this is a bit abstract, consider that those who want to experiment on the human embryo will make all kinds of claims, including the claim that fertilizing a human egg in a glass dish is a cure for infertility; it is not a cure for infertility, it is side-stepping the real problem of infertility and, at the same time, causing irreparable harm to many human embryos.

So it is with the sperm docking in the pore of the ovum, or egg, releasing a wave of calcium irons, a zinc spark, and the closing of the now human embryo's wall, excluding the entry of subsequent sperm and, simultaneously, beginning the independent existence of the human person: the goal of human development being the manifestation of the person from conception.

For, once this initial moment has begun, like lighting a firework fuse, it cannot be taken back, only stamped out; but, if it is not stamped out, then it will inexorably lead to the firework's detonation and the splendour of the manifestation of the human person. At the same time, let us not forget that, just as deep-sea diver has oxygen tanks and gear that enable being submerged, so the human embryonic child, as he or she develops, has the placenta and all that makes living in the womb possible. However, just as the deep-sea diver discards the equipment on emerging from the water, so does the child needs help to discard the umbilical cord and placenta as they are no longer needed. But, given the mentality that calls the conception of a human embryonic child a "blob" of cells, this "blob of cells" is so highly organized, generating its own "interface" with his or her mother, that it is simply untrue to claim that this is anything but a child and the equipment necessary for human growth.

Just, then, as a Catholic sacrament involves an outward material and a word of God, in the ritual which the Church recognizes as coming from Christ, so the pouring of water and the invocation of the name of God the Father, God the Son, and God the Holy Spirit, bring about *Christian Baptism*. At the same time, in the case of human conception, there is an outward change and there is an inward change: the outward change is the coming together of the sperm and ovum and, simultaneously, there is the inward change which is that God has created the whole human being, constituted from what is received from both God Himself and from the child's parents.

Why do we need to draw upon the existence of God?

The question of cause has an ancient origin as, indeed, does the existence of what can be caused. In the case of what exists, it includes the whole universe, the animal and plant kingdom, and men and women, not to mention events, like the Hebrews being brought out of Egypt by an action of God that transcends magic or human effort. In other words, if we truncate what exists to what we redefine as "material events", excluding miracles, we not only presuppose what exists but we also deny the medical evidence of many cures which cannot be accounted for by modern medicine. It is possible, however, like with the existence of the universe, to speculate that a cause as yet unknown can be found. However, like the years that are used to calculate the longevity of the universe, they are so elastic that they can almost mean anything anyone wants them to mean; and, at the end of it, someone can claim but then we came from a prior universe that collapsed. In other words, there are some explanations that do not explain anything.

In the case of a Eucharistic miracle whereby bread, especially the special type of yeast free bread that has been used to make the Catholic, unconsecrated host, there are the usual ingredients of flour and water. Prior to the act of consecration, which changes the whole substance of the bread into the whole mystery of Christ, there is bread and, after the act of consecration, there is the presence of Christ under the appearance of bread. So, while the appearance of bread remains, Christ is now present. A Eucharistic miracle, however, is not the same as this Eucharistic change which, in and of itself, is not verifiable. having said that, a Eucharistic miracle takes place in that through the use of Christ's words at the Last Supper, by an ordained Catholic priest, in the context of the liturgy of the Church in which they are expressed, there is the mysterious change of substance which, while preserving the outward appearance of bread, is now the *Body of Christ*. Similarly, the outward words in which a Catholic marriage is celebrated, express the union of a man and a woman as husband and wife; and, therefore, while

Chapter Three: An Actual Beginning: Conception

nothing has outwardly changed, yet everything has changed for this man and for this woman, who are now joined by God in sacred matrimony. And matrimony is sacred, in this instance, because an act of God has irrevocably changed two baptised Christians, with no impediments and with a good intention to love each other freely, faithfully, for life and, at the same time, are open to the possibility of becoming parents through each other. So, the barest outward sign of words, a ring, or a crown, and the liturgical context transforms a single man and a single woman into a married couple.

In the case of an external, Eucharistic miracle, whereby a consecrated host becomes visibly different, either tinged with red or, as it sometimes happens, becomes flesh, we are conscious of a mysterious causation which goes beyond what can ordinarily happen. This becomes more and more evident when the flesh is examined and turns out to be heart tissue and, what is more, heart tissue that has suffered a trauma. These appearances of Eucharist-as-flesh have been examined by cardiologists who, in general, are going on the evidence of their examinations and not faith. In other words, then, there are phenomena that require a "cause" that exceeds human capabilities; and, while this is often discounted as if there can be a random occurrence of an event that transcends human ingenuity, it is nevertheless evidence in the field of science which goes beyond what is purely scientific.

Similarly, with the ancient question of does the universe have a cause, there are possible explanations which "claim" that this universe is one of many or one in an infinite series of universes. But, in practical terms, these questions are unanswerable in view of the limitations of human investigations. We cannot "see" beyond our present universe although, naturally, there can be endless speculation about whether or not there was a previous universe, multi-verses, or an infinite regress into the past. Arguments about the past or other universes resolve into one: is there an infinite number or not? At the same time, there is a rational argument as to why there cannot be an infinite regress, which is that you cannot have an "infinite regress" and add to it. Therefore there is no infinite regress. If, then, there is no infinite regress, then the past is finite; and, if finite, then it had a beginning; and, if there was a beginning, what caused it? Now a cause has to be greater than the effect. The universe is a phenomenal entity; and, therefore, the cause of the universe is profoundly powerful, which we call God.

How does this help us with the question of the conception of a human person?

The question of the beginning of human life, whether in general from the dawn of time or, subsequently, is the same question: Is there an adequate cause for what comes to exist? We know that human beings come to exist. They do not exist and then they exist. There is a call from non-existence to existence. Nothing can bring itself to exist. So a human being cannot cause his or her existence. So, either an entity has always existed or it is caused to exist. The only kind of entity that makes sense to exist, per se, is God; and, not just any kind of God, but a God which is relational: both in the mystery of being three persons

Chapter Three: An Actual Beginning: Conception 41

in one God and being capable, therefore, of creating all that is caused to exist, to exist-in-relationship.

What, however, makes the biological cause of human being's existence an inadequate cause of human existence? In principle, what is biological causes what is biological to exist. A group of bacteria causes bacteria to exist. A group of amoeba cause an amoeba to exist. A male and female dog cause a dog to exist. So why cannot a male sperm and a female egg or ovum cause a biological entity to exist? Actually, when human conception does not take place, there is a quasi-biological entity that comes to exist, namely a hydatidiform mole: a kind of semi-amorphous growth that is not of itself ordered to both being and becoming a human being.

Unless an argument assumes that all that exists is of a uniform nature, such that everything has the same characteristics, then it cannot follow that a chemical "fathers" a bacteria, that a bacteria "fathers" an animal, that an animal "fathers" a human being. But chemicals, bacteria, animals, and human beings do not share all the characteristics typical of each type of creature. On the contrary, the human being takes up all that is characteristic of other types of creature but then expresses what is not found in them, namely, a personal being and all that that entails in terms of the pursuit of meaning, relationships, love and the impossible to describe variety of human creativity. In other words, without the assumption of uniformity, it is not possible that one kind of entity or creature can lead to another one which is more developed than itself; for, in the end, the goal of all creatures, namely, being human, would have to be present in the beginning. The evidence, clearly, is that human intelligence and ingenuity are not present in either matter or the earliest forms of animated life. What is present, however, are the capacities of organization, of intelligence, reproduction, movement, and making or finding a dwelling, which correspond to the "level" of the biologically material, creaturely existence that it possesses.

When it comes to found objects, like a piece of wood resembling a face, a stone shaped like an axe head, or a fossilized shell impacted or cast on a piece of stone, the agent-actor, as it were, is the human being who recognizes an echo of a human face, a shell out of place, or a tool. In other words, there is consciousness and communication, education, and the grasp of an object's meaning. Similarly with the design of electronic devices, medicines, and the composition of paintings, music, and writing, the agent-actor is the human person. So, clearly, the human being is capable of taking and shaping materials, drawing on their natural properties, taking components made by others and generating a design, a method of making it, and the problem-solving necessary to completing what was begun.

Why God is a necessary cause of human existence

The principle, then, of an adequate cause of what has come to exist, entails really thinking about what has come to exist. A book has come to exist and the author or illustrator, along with the publisher and production company, are an adequate cause of what exists. Clearly, in this case, there is a clear transcendency of the author and publisher over the materials of paper and cardboard. Even if a computer writes a book, it only writes what the computer engineers have made it possible for it to write. What is more important, the computer has no consciousness of what it has done. In other words, whether a book is produced directly, by an author, or indirectly by a machine made by human beings to make books, the inanimate author transcends the materials to the extent that the one creates with the other: that the one is like clay in the hands of a potter.

Similarly, then, unless we suppose that a box, a washing machine, or a motor car builds itself, there is an agent-actor who, even if using robots that have been built for the purpose, is at work. And, therefore, there is a multi-level difference between the materials that are made into an object and the person who designs and does the building. Just, then, as the gardener makes use of the seeds that grow, separating them out and transplanting them, so there is a significant difference between the gardener and the plants. And so on with the animal kingdom, whether the animals are allowed to run free in a reserve, are farmed domestically or are trainable pets, it is the man or woman who trains, harnesses, and makes provision for the welfare of the animals.

When we consider, then, the coming into existence of a human being, a child, we are again confronted with this chasm of causation. On the one hand there is the contribution of the sperm and the ovum or egg which, on contact, begins an irreversible process of manifesting the person present from conception. On the other hand, as this exploration has striven to show, what emerges from conception, namely, the human person, exceeds the biological contribution to conception, just as writing a book exceeds paper and pen, ignition the match and candle, and the work of the gardener exceeds the existence of plants. So it is not a matter of invoking God, as if there is neither a credible argument for God's existence nor an intimation of His power; rather, it a question of the existence of an uncaused cause being the only rational explanation for the coming into existence of the human person.

What is more, owing to the Christian Revelation that God is *three person in One God* and is, therefore, an interpersonal being, so it makes even more sense that there is a personal cause, of a personal, interpersonal being, made in the image and likeness of God. And, as such, just as the Blessed Trinity is the primary cause of the personal existence of each one of us, so this is expressed in the contributing, personal cause, of the husband and wife.

Chapter Four

The Frozen Embryo Debate and *in-vitro* fertilization (IVF)

Fertilizing a human egg with a human sperm in a glass dish translates an interpersonal act into a product. However, the person who comes to exist, who has never existed before and will never exist again, is not a product – but a person. So while the context does not change the event, namely, of a person coming to exist, it changes the perception of that event in the eyes of a technician who has brought it about. Between the process of obtaining the human eggs, including the overstimulation of the woman's ovary, the obtaining of sperm from the man by self-stimulation, and the arrangement of numerous acts of fertilization so that there are a number of human, embryonic children, there are any number of human casualties. The possibility of multiple parents: from biologically derived human egg and sperm to another woman carrying the child, to other parents waiting to see if they want the child.

Then follows the so-called quality testing of each of them, in the course of their early development, all of which leads to a rejection of some of them. There are those that are discarded, for whatever reason. There are those to be experimented upon. There are those to be implanted and then there are those that, once implanted, are "reduced" by destroying them. Then there are those to be frozen. Although the numbers of frozen embryos are running into the millions, there is a tragic loss of life as they are thawed. Hannah Strege, the first frozen child to be thawed, implanted, and brought to term, had some twenty or so siblings who did not make the transition from frozen to thawed.

What was argued to be a "cure" of fertility, of a woman with blocked fallopian tubes, the tubes which carry the ovum or egg down to the place of fertilization and then on to the womb, was not a cure of infertility at all. Whatever the cause of blocked fallopian tubes, there does not seem to be been attempted any actual help to remedy that condition; rather, profiteering being what it is, much was made of the emotive claim that IVF "cured" infertility: IVF cured nothing. Indeed, all the technological skill that was developed to circumvent helping the woman's actual cause of infertility, could be employed to actually help with discovering the real cause or causes of infertility. What is more, there could be so much more done for what is often both a difficult and traumatic event, namely a pregnancy that has implanted in the fallopian tube, known as an ectopic pregnancy. But, while this is a real need, there is not very often the medical will to investigate and help women with what actually the real problems are, although there is one university doctor-researcher who is currently investigating how to help women with an ectopic pregnancy. Although, having said that, there is a lot of help from those who are very familiar and well versed in observing the fertility-infertility cycle of the woman.

It also happens that, when a couple refuse the manipulation of human life and adopt children who need help, whether children who are already born or adopted, embryonically, that an otherwise infertile couple conceive. We do not know all the answers to the many questions that exist but more could be known. But, let us not forget, a child who comes to exist naturally wants to know the history of his or her existence and family; and, therefore, it is a great injustice to the child, in addition to everything else, to obscure the "coming-to-exist" through marriage which illuminates the identity of a child.

Questions that need to be asked and answers that need to be given

There is an immense amount of money to be made from those who market the conception of children. At the same time, there is an immense contradiction between claiming to help a person have a child and destroying children in the process. At what point, then, does it need to be asked: What are the causes of infertility, either in the male or the female?

Impotence is not necessarily a physical condition. According to those who study the use of pornography there is a condition whereby addiction to pornography makes it impossible for the man to be aroused by an actual woman; and, if this woman is his wife, then this is increasingly a cause of divorce. This in itself raises a variety of questions concerning the development of pornographic addiction, concealing this addiction from a fiancé or wife, or a general denial of its seriousness. Another concern, although difficult to quantify, was the absorption of lead by road runners and its damaging effect on male fertility; but, with the removal of lead from petrol, there still seems to be a concern about lead and other causes of male loss of fertility. Atrazine, a common pesticide, while it is principally a cause of feminization in frogs can also cause a loss of fertility in men. In other words, men are susceptible to a loss of fertility because of the effects of various chemicals or, in one case, the impact of lead on male fertility.

Similarly, there is a massive number of hormonal pills being poured into the woman's body and, once through their system, into the general water supply. This chemicalization of the body has more ramifications than just male or female pollution as it expresses a general willingness to administer unnecessary medical treatments leading, generally, to a flattening of a woman's sexual drive and many, many, side-effects, leading to more chemicalization of the woman's body. Then there are unnatural objects inserted into the woman's body so that, in general, the whole focus is on preventing the conception of a child by polluting or interfering with the woman's body. Then there is, again, the further chemicalization of the woman's body by abortion pills, and all their side-effects, as well as the dangers and consequences of surgical abortions. All of which damages the woman's relationship to reality, to the child, to the place where it took place and indeed to her own body. In other words, and in so many studies of women's health, the objective seems to be the reduction of fertility and not the discernment of what will enable a man's wife to conceive. In-vitro technology, then, is a contradiction in terms. It arises out of not treating infertility, particularly the woman's infertility, and costs immeasurably more than coming off unnecessary drugs and learning to read the signs of fertility or, sometimes, good observations can lead to a cure, actually, of the actual cause of male or female infertility.

As regards the poor human embryo, which has become commodified, along with trafficking human beings, we have entered a new slave culture whereby the human person is not an "end in him or herself" but an object of commerce. Or, where it is possible to choose the sex of the child, very often it is the girl that is rejected and discarded even if, as in some countries, this has led to a devastating imbalance in the number of men and women in a society.

The present situation of a frozen human embryo

The IVF industry, then, has created a multifaceted way of investigating and destroying human embryos which, by fronting the reality of what it does with an emotional appeal centred on helping people who are infertile. So, while it is true that there are many children conceived and brought to birth, the full force of what is happening is hidden. There are multiple injustices: to the child conceived outside of the marital act, there is an uncertain outcome; to the child failing the quality control check and being discarded, the outcome is no longer uncertain; to the child experimented upon, the outcome is no longer uncertain; to the child "deselected", once implanted, and no longer wanted, the outcome is no longer uncertain; to the child aborted, because of a real or an imaginary imperfection, the outcome is no longer uncertain; to the child frozen, there is still an uncertain outcome; and to the woman and the man who think that this is a treatment of infertility, it is not, because the cause of infertility is not cured. Whatever the real infertility is, IVF is not a treatment of infertility; it is a manipulation of the beginning of life to make money. As regards the human embryo chosen for freezing, the natural liquid within the embryo is removed and an anti-freeze type liquid is added; and, using liquid nitrogen, the human embryonic child is fast-frozen to prevent damaging ice forming in the child. And, ultimately, these children remain in the freezer until they are wanted or, alternatively, until an unknown date. Children have been known to survive thirty years in the freezer; but, obviously, they cannot last forever although some are planned to live up to 55 years in a frozen state and may live longer. However, occasionally, the equipment fails, and the frozen embryonic children die.

So now there are millions of frozen children. If there was any uncertainty about whether what is frozen, is a child then money makers would not be providing human embryonic children at a price. Thus, all the children who are born to this procedure, are not in themselves products, but children. And a child is a gift from God. But after does not justify before. So the fact that children have come to exist does not justify the method through which they came to exist. Nevertheless, the child is loved by an everlasting love, namely, God; and, therefore, the question for the servants of the Lord or of any good intentioned person is: How to overcome the completely frustrated human development of this child and prevent this happening to other children? However, there is an ethical problem which is how to help the remaining frozen children come to term and to be born. Some think that the answer is to thaw the embryonic human children, baptize them, and let them die. Some think we cannot do anything, as the immorality which led to their existence leaves us in an impossible situation of having to use

Chapter Four: The Frozen Embryo Debate and IVF

immoral means to rescue them, namely, that they will be transferred to the womb of a woman with a pipette or equivalent instrument, thus continuing to employ the illicit means by which they have come into existence and are "manipulated", prior to being used, or misused.

By contrast, however, helping a frozen human embryo to complete the frustrated human development, which is a totally imposed frustration, for which the child bears no responsibility whatsoever, is about overcoming the human irresponsibility that brought about the child's predicament. So, a person is tied up and thrown into a lake and is drowning. The rescuer, rowing out in a boat, cuts the cords that bind the person drowning, but then uses the rope to secure them to the boat as the man in the water is too weak to swim to the shore. Alternatively, a man has been thrown into the sea in a punctured plastic bag; but, being rescued, the rescuer turns the plastic bag into a float to keep the injured man alive while the rescue is completed. Similarly, the frozen embryo is stranded in liquid nitrogen by a technician and cannot rescue himself. Thus the rescuer, unfreezing the human embryonic child, takes the same instrument, the pipette, that was used to put the human embryo from the glass dish where he or she came to exist into the freezer, to make the transfer of the human embryonic child into the womb of the adopting mother. In other words, without condoning how the child came to exist, outside of the womb, the same instrumentation is used to rescue the child from arrested personal development: the development which will, over time, reveal the presence of a child whose life has been so totally frustrated.

Is there a theological justification for helping the frozen child?

In the widest possible sense, each of us is called to love like Christ; and Christ loves us all. But, more than that, at His Incarnation, He united Himself in a certain way to each one of us; and, therefore, as each one of us comes into existence, so each one of us comes into existence-in-relationship. We exist-in-relationship to God who caused us to exist and to His Son, Jesus Christ, who elected to be united to each one of us through His Incarnation. Each one of us, beginning with Mary, is given the gift of the Holy Spirit; however, as in her case, and the case of Adam and Eve, they were given the gift of the Holy Spirit at conception. Adam and Eve, however did not recognize and live the relationship of being whole, and holy – but ruptured it. Nevertheless, God prepared Mary, and Mary accepted the preparation, for the coming of our Savior, Jesus Christ, to be born of the Virgin Mary and to cross, through the crucifixion, the divide between us and God. And, in that each of us is united with Jesus Christ in a certain way through the Incarnation, the Son of God dwelling in human flesh as true man and true God, His resurrection became the foundation of ours. Thus, having founded our salvation on what happens to Jesus Christ, our Savior, God restores our ruptured relationship through adopting us in Baptism.

The divine act of coming into contact with each one of us, through the Incarnation and, subsequently, taking our relationship to Him through His life, death and resurrection, founded our salvation on an indiscriminate act of adoption: that we are all, both individually and as a human race, adopted into the possibility of salvation. Similarly, when it comes to the frozen human embryo, no one is discriminated against: each and every frozen embryo has the inalienable human right to continue the development that was, literally, and unjustly, put on ice.

As regards the frozen embryo which, if it thaws, implants and matures successfully, it is entitled to the wholehearted welcome of parents who want the opportunity to love the humanly forsaken child. And, while recognizing the difference between St. Joseph and an ordinary human father, the welcome of that child is equally an expression of the practical love of the father as it is of the mother. And, if there is a Scriptural figure who epitomizes the mentality of adoption it is St. Joseph; however, any welcome of a chid not our own, involves welcoming that child, remembering that whether the child is conceived in rape, through adultery, or by IVF – the child is not responsible for what other people have done.

So, in sum, the theological argument for the human adoption of frozen embryos is based on the divine adoption of each one of us, through the Incarnation of the Son of God in human flesh.

Chapter Five

Identity; Transgenderism; and the United Nations Bureaucrats

The human question of identity has been with us from time immemorial and has entailed both constancy, change, and indeed development. While our nature remains as it is, both fallen and falling within the reach of salvation history, we are present to any number of challenges to what is entailed in being a human person. And, let us remember, just as the concept 'person' had its roots in Greek drama and was then taken up to account for the nature of the Blessed Trinity: the being-in-relationship. There is God the Father, God the Son, and God the Holy Spirit, who are not three gods but one God. So the concept of being a human person expresses, profoundly, that each one of us is a human being-in-relationship while, at the same time, being one human race and not many. And, therefore, we are made in the image and likeness of God.

Growing up is growing through

Each one of us, however, comes into consciousness as the physiological basis of psychological development unfolds. In other words, the physiological basis of consciousness is not that physiology is consciousness but that, as hearing develops, so I hear sounds and speech; and, as I get older, so I hear words and concepts which make me think. Thus, as the baby has enough bodily development, including hearing, seeing, touching, to engage, so does smiling, giggling, laughing, and the whole range of infant to toddler to active pursuits start to show themselves. Words, vocabulary, reading, drawing, and whatever else, all of which start to show themselves, along with friendships, that we are both an individual whose identity is discoverable, but also a being-in-relationship who has a mother and a father and a whole host of unfolding relationships.

There are, of course, specific questions which are uniquely our own, although they have a widespread existence, such as what are my talents, what kind of work do I want to do, whether to marry or not, or to become a monk, a nun or a priest. Or, is there a God? Am I religious? What is the point of my life? But, in and amongst these questions, there can be what seem, at the time, almost intractable questions about our particular experiences which, going on the many social problems that are running through our times, can range across any number of problems. Whether it is running away, suicide, suicidal ideas, self-harm, abuse, trafficking, anorexia, being overweight and disliking our bodies, wanting to change our facial structure with cosmetic surgery, to be like an "ideal" westerner, to have a white face, drugs, gangs, assaults, alcohol, spiked drinks, pornography, violent videos, excessive screen time and loss of the outdoors, compulsive checking of "likes" and "dislikes" and on-line contacts instead of in-person meetings, with all the vulnerability that that entails. War and its carnage, money manipulations and machinations, casualties and the utter indifference to peace while war mongers are either hidden or have palaces. Each and everyone of these real problems is a serious event with all its attendant problems; and, at the same time, there are those who, in different organizations, are seeking to genuinely help and rescue, heal and help recover distorted and disfigured lives.

So, it is natural and inevitable, in a sense, that young people are questioning what exists. However, what is unnatural and questionable is when young people do not question what other people say, especially when it comes to their own lives and welfare. But, more widely, it is indeed a part of growing up to really investigate what goes on, why we have been through what we have, where are our lives going? It is not healthy to dive into ideas, especially those driven by people who do not want dialogue, and to be immersed in an uncritical environment – especially one that wants to isolate a person more and more and propose more and more extreme solutions to what, in fact, is the time to discuss, to discover, to go out and about and find out about life.

Can I be in the wrong body?

Generally, in the course of growing up we can go through all kinds of ideas about ourselves, ranging from wanting to be a footballer, astronaut or scientist and struggling to be a writer. But there can also be questions about our sex and, for myself, I can remember thinking it would be easier to be a woman, rather than being a man, because I would not have to earn a living. Thus I was neither realistic about myself nor about what it is to be a woman. And, in the end, it was an attempt to escape the questions of what to do with my life; and, indeed, after a while,

I even forgot I thought this and started working as a journey man, doing basic decorating and building work. However, the suffering did not disappear, as I neither wanted to be a journey man – but I did not know what else to do; and, what is more, I hated handling the "poor payers": the people that took on a job of decorating and then said how little money was available for it.

Now there is the claim that a person can be in the wrong body. But while this is just a claim, impossible to prove, it nevertheless depends on some kind of idea that a body exists and a personal identity somehow comes to inhabit it. But to be "inhabiting" the wrong body, as if the personal identity descended from a prior existing place, there would have to be some kind of mechanism by which a change occurred that what would have normally been a boy ended up as a girl. Thus a "girl's identity" inhabits a boy's body. The problem with this is that there is no possible observation that can confirm this theory. Moreover, it supposes that a human person is not a whole, from the beginning, and does not unfold outwardly what is inwardly determined by what happens at conception. Alternatively, a boy develops a fiction, as it were, that he is a girl and, as such, that fiction inhabits a boy's body. In either case, however, there is this "disconnect" between an "identity" and a bodily expression of it. Or, again, more as a result of persuasion, a child is convinced by others that the reason for any disquiet that the child has about his or her bodily being is "explained" by being in the wrong body. The problem with this explanation is that it generally runs with "Do not discuss this with your family or any other person, outside of "us", insist on special pronouns, and be prepared that the solution is the chemicalization of the body and surgical mutilation". But the chemicalization of the body and surgical mutilation is not called or described in this way, nor is it made clear that all this can lead to life-long medical alterations that no more change a person's sex than wearing the clothes of the opposite sex. In all these instances, there are all kinds of possible psychological flaws or impediments in a person's growing up, from rejecting femininity or masculinity because of trauma, witnessing abuse, seeing the exploitation of women or the rejection of men, or just being drawn into confusion through peer pressure, social media, and influencers.

Unfortunately, however, our society has progressed from using unnecessary chemical products on women, called contraceptives of one kind or another, to pharmaceutical products that unnecessarily kill an unborn child or, through unnecessary surgery, the child is unnecessarily mutilated and taken from the mother or the woman carrying the child. In other words, there is an immense amount of money and technological investment in unnecessary chemicalization of the body and changes to it. If this, then, is followed through there is a tragic outcome of a child or young person being funnelled into the puberty blockers' chemicalization of the body, with its attendant risks of diminished bone density and infertility and, in the end, surgical mutilation to make certain parts of the body mimic the outer parts of the opposite sex.

At the same time, these alterations of human bodies serve no practical or medical purpose and render a person subject to chemicals and surgeries indefinitely, making money for those who carry out these procedures and ensuring a mentality of widespread confusion and infertility. What is more, when a young person begins to understand what is happening to them and wants to withdraw from the whole charade, he or she does not want to see the counsellors who pointed them in this direction, ignoring all other indicators about his or her psychological state. Furthermore, there is a widespread lack of accountability in that the clinics who advanced all these kinds of procedures did not even keep records of who they saw and what the outcome was, suggesting that such a lack of accountability is an indirect way of avoiding liability for what has been done to the young person. In other words, the whole "system" seems designed for the exploitation of young people and not for their good.

Ideology

What, then, defines an idea as an "ideology"? On the one hand, it could be said that "idea" and "ology" make up the study of ideas; and, in one sense, this is true. But there are two kinds of studying, one is about looking at the idea, rather like a plant, and asking where does it come from, what is the soil it grows in, and what will it unfold into. This is a living study, going back to the origin of words, early texts, whether in translation or in their earliest languages, albeit full of difficulties. Like the idea of the "soul", or psyche, being like a captain in a ship. Or, alternatively, the soul is that which determines the inside of what the outside of an object is; but, just as the inside and the outside of a box are inseparable from the shape of the box, so the soul is naturally inseparable from the body: as the human soul is what

Chapter Five: Identity; Transgenderism; and the UN 61

determines a human body to be both living and a whole human person. Or drawing on Hebrew, instead of Greek ideas, and discovering that the human being is a "whole", although describable in terms of assembling parts, nevertheless is an organic whole which is given life and lives.

On the other hand, there is the study of ideas which is more about leading people into a room, dropping a lock on it, and leaving them in a locked room, windowless, and saying that this is the real world. If people are inexperienced, not very well educated, unused to critical thinking, never watched or planted real plants and watched them grow, seeing both disappointment and disease, then people are vulnerable to the belief that the real world is a virtual world where anything and everything can happen at the click of a button or a line of typed code. But there is also an immense world of culture to be drawn upon, rejoicing in the success of a plant growing and building on it, slowly, over many years, both studying what and how it grows and what helps and alerts there are to problems. Not to mention the benefit of talking to those with more experience; but, all the while, watching, trying things out, changing methods, and noting what is good, helpful and lasting.

Ideology, then, in this more limited and restricted sense of building a room of one idea, has numerous takes, including transgenderism. This is because falling into a group chat which excludes those critical of it, forbidding discussion, swallowing special words, believing the paintings of paradise on the wall are wonderful and true to what is "out there" are all signs of a profoundly enclosing "in-think". This is especially confirmed by relying on the law to restrict discussion, questioning, challenge, as if by way of saying that this "idea" is above being proved, demonstrated, corroborated by evidence, so that any dissent from it is immediately bigotry, prejudice, malice and not, simply, the outcome of questioning what is questionable. Furthermore, those who advocate ideologies, like transgenderism, on seeing the harm it has done to young people, families, law making, the keeping of social statistics, do not regret the confusion, the surgeries, the dependence on the chemicalization of the body, long term, if not for life. Rather, the advocates of this "isolation ward" profess injustice, unjust scrutiny of what are, in effect, self-fulfilling claims that are contradicted at every turn, especially by rendering healthy young people unhealthily mutilated.

So, an organization has brought together a number of countries in the name of common concerns, and agreed, for instance, that there are two sexes, male and female, as in 1995 at Beijing. Now, the bureaucracy of that social institution is proclaiming a "closed shop": a darkened room: a place of no-go dialogue. In other words, contrary to the original purpose, founding documentation, and spirit of the founders of the United Nations, a group mentality has all but colonized this institution and rendered the people in it almost completely incapable of dialogue. Nothing is advanced by argument and everything is advanced by "name calling". In other words, those who argue for the true health of women, the unborn, young people, are not argued with so much as rubbished, being called "anti-rights" and enemies of the development of women. In reality, however, the proper study of non-medical methods of helping people, both women generally and husbands and wives, through natural family planning, waiting for marriage at a reasonable age, having children, and educating them in the wisdom of being healthy is overridden by wealthy promoters of anti-population policies.

What we need to recall, as simple as it is, is the foundation of human action: do good and avoid harm.

A Word About Free Speech

Clearly, there cannot be free speech if one argument dominates the social space, whether it is one political argument, one theory of medicine, one account of human existence, religious experience, or the nature of human being. What, then, is the prevailing mentality of those who dominate a social space, like journalists, bloggers, website controllers, social media gate keepers, government officials, biotechnological company directors, abortionists, and the wealthy manipulators of public opinion, whether openly or through subterfuge and agent-actors? Do we live in a subtly but really "controlled environment"? Is that where those with the where-with-all are going?

When a child in the womb is claimed to be a "blob of cells", who wins but the so-called abortion industry that can clean out and sell those cells because, after all, they are the cells of a human being? When people question when life begins and, at the same time, market products to end it, to chemicalize the bodies of women, who wins but the marketers who change and vary products that destroy life and bodily health? Who has agreed to the free speech of the child in the womb being silenced? Who has agreed that body language is no longer free speech: that growing in the womb is no longer life and development – but then go on to sell and market the baby manufactured in a laboratory, quashing into the eternal silence the lives of those discarded, experimented upon or simply trashed or frozen?

So, when a person peacefully protests the injustice to the unborn of abortion, the horror of what happens to the child, the help that a mother needs, whether coerced or not into aborting her child, and this protest is criminalized, then we are losing the freedom of the social space to those whose power can govern its use. If, then, a woman can be taken to court because she is a woman and a man, whatever he says, is invading her space and she objects, then the fencing in of the social space is clearly more visibly present than was realized; and, what is more, is more pervasive than was realized. If a sport which is intended for each sex, separately, to preserve fairness in competition is turned into a male sex competition because men are claiming to be women and are supported by the law to do so, then not only are men competing with men but men are unfairly taking advantage of stealing women's records, achievements, and awards. Not to mention the medical confusion of recording a male as a female, a man as a woman on passports, in safe-spaces for women, in protected environments for children, then who is controlling the social space? If teachers, doctors, nurses, clinicians, surgeons, lawmakers, judges, police and any contributor to the social life of a country or a place cannot speak objectively concerning the well-being of those in their care, then they have lost the credibility in their professions and, again, are fencing in the social space.

Perhaps, then, we are not talking about the existence of free speech so much as the freeing of speech from those who control it for those who need it.

Having unmoored the truth from reality, we will no doubt be like collectors of splinters, splinters of the true cross which, we hope, will multiply like the Eucharist and build a new ark of civilization.

www.ingramcontent.com/pod-product-compliance
Lightning Source LLC
Chambersburg PA
CBHW060852050426
42453CB00008B/958